MAISIE MAMMOTH'S MEMOIRS

A GUIDE TO ICE AGE CELEBS!

HELLO, I'M MAISIE

This is my memoir! Would you like to meet some
of the biggest stars of the Ice Age? I was a pretty
big name back then and I know all the juiciest
secrets, attended the most star-studded events,
and witnessed the most villainous deeds.
Trust me, you won't want to miss this.

MAISIE MAMMOTH'S MEMOIRS

A GUIDE TO ICE AGE CELEBS!

Illustrated by Rob Hodgson
Ice Age expert Prof. Mike Benton

Thames & Hudson

CONTENTS

WOOLLY MAMMOTH
Maisie Mammoth
P.8

DIRE WOLF
The Brat Pack
P.28

YUKON HORSE
Betty
P.29

SABRE-TOOTHED CAT
Stella the Star
P.10

IRISH DEER
Conor
P.38

GIANT BEAVER
Bill & Ben
P.40

SABRE-TOOTHED SALMON
Jackson
P.20

GIANT ECHIDNA
Matilda
P.34

DOEDICURUS
Sir Spike-a-lot
P.18

GIANT TERATORN
Paula

WOOLLY RHINO
Hector

GIANT SLOTH
Laid-back Luis

GIGANTOPITHECUS
Gavin the
Giant Ape

GIANT POLAR BEAR
Ursula

STEPPE BISON
Sadie

ARCTIC GROUND SQUIRRELS
The Furry Fan Club

GIANT SHORT-FACED BEAR
Maria

TITANOBOA
Victor the Villain

WHY WAS THE ICE AGE SO COLD?

Our Ice Age started about 2.6 million years ago and scientists call it the Pleistocene Epoch. It happened when our planet's temperature went down – and stayed down – for a very long time. Giant fields of ice formed across the Northern Hemisphere and snow covered the land all year round. (It certainly got a bit nippy for us mammoths.) Brrr!

OUTER SPACE
Changes in how the Earth moved around the sun might help explain why we all got so chilly.

SLOW MOVERS
Ice began to form around land that had drifted close to the North Pole. The ice fields reflected the sun's rays, sending heat back into space, which allowed the surface to cool further – forming even more ice.

SOMETHING IN THE AIR
There was some seriously scary volcanic action and greenhouse gas levels changed. The energy from the sun changed too.

SNAP!
Our fabulous Ice Age never ended – it's just reached a warmer stage, which scientists call the Holocene Epoch. You're an Ice Age star, just like me!

HOW TO DEFROST A WOOLLY MAMMOTH

40,000 years ago, my unlucky personal assistant Buttercup tripped over her own feet and fell into a bog. The freezing temperatures kept her perfectly preserved. Buttercup always wanted to be famous, just like me. Now, scientists all over the world are super-excited to meet her! Here's how they do it.

1. TAKE ONE FROZEN MAMMOTH

Poor old Buttercup has been in the ground for a long time, so this has to be done carefully. She's the biggest ice cube ever!

2. STAY COOL

To make sure that Buttercup isn't damaged, she has to be defrosted slowly and gently. Keep your super-sized ice cube in a nice cold ice cellar until you are ready to defrost it.

3. USE GENTLE HEAT

Some scientists have used racks of hairdryers to slowly warm things up.

4. KEEP YOUR EYES PEELED

Watch out for bits of plants that might have been frozen with the body. They could teach you amazing new things about Buttercup's busy lifestyle.

WOOLLY MAMMOTH

MAISIE MAMMOTH

NORTHERN HEMISPHERE

We've already been introduced, but I'm Maisie and I'll be
your Ice Age tour guide. I've travelled all across the Northern
Hemisphere so I'd like to think I'm pretty qualified for the job.
Sure, I got into a few pickles along the way. (Neanderthals*
enjoyed mammoth meat a bit too much for my liking.)
But I'm a survivor and I was always up for adventure.

With a layer of fine fur close to my skin to trap warm air and a
rough strawberry blond coat over the top to keep me dry, I was
prepared for anything. But don't be confused if my herd doesn't
look like me – we were all different colours! My cousin Emily
was famous for her glossy chestnut locks and Aunty Boo's
dark coat made it easy for her to melt into the shadows.

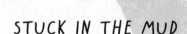

STUCK IN THE MUD

The reason you have a good idea of what I looked like is because
some of my lucky relatives were found recently, perfectly
preserved in ice. In fact, a few greedy-guts still had food in their
stomachs. In 2012, a boy found a mammoth while he was walking
his dog. Who'll be next? (You never know, it could be you.)

FLOWERY FEASTS

Mammoths like me found food by using our long, curved tusks to scrape snow off the ground. Mine were 4 m long! My family were rather partial to wildflowers. (The plants were packed full of protein!) Buttercups were my favourite, which I picked using my trunk.

WHY IS A MAMMOTH LIKE A TREE?

No, it's not a joke. My tusks were as thick as tree trunks and you could find out my age by counting the growth rings inside them – just like counting the rings of a tree.

BACK FROM THE DEAD

Would you like to bring a woolly mammoth to life? Some clever scientists are hoping to try out their cloning skills with the help of my DNA and my closest-living relation, the Asian elephant. Things may be about to get hairy!

*To find out about the Neanderthals, turn to page 42.

9

SABRE-TOOTHED CAT
STELLA THE STAR
USA

Once seen, never forgotten! Stella's trademark was her megawatt smile and with canines that measured up to 28 cm (that's almost the height of the book you are reading now!) she was very hard to miss. If you thought that her roar was worse than her bite, you'd be dead wrong.

ALL YOU CAN EAT
Stella's favourite joke was that she liked the woolly rhinos but she couldn't eat a whole one. She mostly dined on deer, bison and camels.

Rumour has it that Stella was always on the lookout for a good dentist. Much to her embarrassment her pearly whites were rather brittle and prone to breaking! For some strange reason the dentists never seemed to stick around...

A MIGHTY MOUTHFUL

Like a snake, Stella could open her mouth extra wide to deliver a quick, powerful stabbing bite to the throat of whomever she was about to eat.

FELINE FATALE

With her thick neck and short legs, Stella might not have been the most elegant of felines, but her stocky frame was perfect for stalking prey without being seen.

GIGANTOPITHECUS
GAVIN THE GIANT APE

CHINA

Gavin never looked for fame and fortune, but fame found him. Gavin and his peace-loving gang only wanted to live a quiet life in the Himalayan mountains and get back to nature. Unfortunately, the tale of the Yeti started and suddenly everyone wanted to catch a glimpse of poor old Gavin, thinking it was him. He tolerated the pilgrimages fairly well – just don't ask him for an autograph!

MOUNTAIN HERMIT
Gavin enjoyed the simple pleasures of life and food was his favourite thing. Like me, he needed a lot of it! Between meals of seeds, fruit and bamboo, he spent his days sitting and meditating, or ambling around on all fours.

BIG DREAMS

When Gavin stood up, he was over 3 m tall and he weighed more than 500 kg. He dreamed of climbing trees like some of his smaller relatives, but he was just too big. That didn't stop him trying though and you could always tell where he had been from the trail of broken trees he left in his wake.

MISTAKEN IDENTITY

As well as being mistaken for the Yeti, people thought Gavin might be a dragon when his teeth were first discovered in China! I'm glad he wasn't around to hear about that.

GIANT POLAR BEAR

URSULA

ARCTIC

Ursula loved to sing. Ok, I admit it sounded a lot like roaring and howling to the rest of us. But would you want to upset someone who is 1.8 m tall on all fours, over 3.6 m long, weighs over 1.1 tonnes and holds the record for being the largest meat-eating land mammal ever? If Ursula said it was singing, then we all agreed with her!

Like all music stars, Ursula had an image to maintain. She always wore a thick, white coat, which was oily to repel water, so that she could camouflage into her snowy environment until she was ready to surprise a fan... or a meal.

RELOCATING
You won't believe this, but giant polar bears didn't actually come from the Arctic. They started out in England and slowly moved farther and farther north, adapting to the harsh environment little by little.

SWIMMING STYLE

As a keen performer Ursula needed to keep fit. Every morning she braved the icy waters for a swim, but don't worry, her thick layer of body fat helped to keep her warm and her large, webbed paws worked like paddles in the water to keep her moving quickly.

MEAT AND GREET

Giant polar bears like Ursula needed to eat a lot of meat. As they got bigger over time, so did their prey! Sometimes Ursula scavenged for food by chasing other predators away from their kill, but not always, and like any of us, she was at her best on a full tummy!

TITANOBOA
VICTOR THE VILLAIN
COLOMBIA

I never met Victor but rumours of his dastardly deeds reached us all the way from his home in the Cerrejón basin floodplain. Victor was the bad boy of the Ice Age – one of those villains that everyone loved to hate.

It all started when Victor tried to impress someone with a party trick that involved swallowing a croc in one gulp. He got a taste for it after that and had crocodile for breakfast, lunch and dinner, with a few fishy snacks in between.

SLIPPERY SURPRISE

Victor could blend in perfectly with his surroundings. One of his favourite hobbies was seeing how close he could slither to his victims without being seen. He loved nothing more than to spend his days half-submerged in the shallow waters of the Amazon basin, plotting his next move.

HISS-TORY LESSON

None of us blamed Victor for being a cold-blooded killer. It was just how he was made. But his fame was short-lived... The rapidly cooling temperature of the Ice Age meant bad news for reptiles like Victor!

SUPERSIZED

Titanoboas like Victor were the largest snakes that ever slithered. They could grow to 13 m long (that's longer than a bus) and weighed as much as 1.3 tonnes.

DOEDICURUS

SIR SPIKE-A-LOT

BOLIVIA

For a lifelong herbivore, grumpy Sir Spike-a-lot
picked a surprising number of fights. He was the
worst-tempered plant-eater in South America
and his top weapon was his stiff, heavy tail.

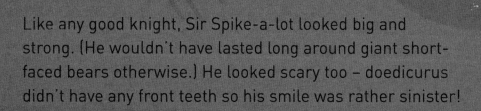

Like any good knight, Sir Spike-a-lot looked big and
strong. (He wouldn't have lasted long around giant short-
faced bears otherwise.) He looked scary too – doedicurus
didn't have any front teeth so his smile was rather sinister!

PERFECT PROTECTION

No man-made shield could match up to a doedicurus' tough shell. It was made from small, linking pieces of bone and it was incredibly flexible. Their shells have been found almost intact over 2 million years later.

READY FOR ANYTHING

Like camels, a doedicurus had a fatty food store on its back, under its shell, which gave an extra layer of warmth in the bitter cold.

A KNIGHT'S TAIL

At 4 m long and weighing 2 tonnes, Sir Spike-a-lot won all his battles claws down. He swung his spiky, club-like tail from side to side to bash his rivals.

SABRE-TOOTHED SALMON

JACKSON
USA

Teen fish-throb Jackson became the hero of his school when he figured out how to swim upstream to reach the gravel beds where fish could lay their eggs. When fish reached a waterfall they assumed it was a dead end – till Jackson started leaping! He cleared the top of the waterfall and others soon followed, inspired by his bravery. To this day, salmon still leap up waterfalls to honour Jackson.

DADDY DAY CARE

Just like modern-day salmon, the male sabre-toothed salmon would defend its eggs. Jackson's spiky teeth would have come in handy if another male got too close.

FANGS-A-LOT

The thought of Jackson used to send shivers down my tusks – until I met him! His fearsome fangs and his size gave him a bad reputation – he was 2 m long and 177 kg – but he was a chilled-out surfer dude who just wanted to ride the waves rather than terrorise the ocean.

CHILLING OUT

When he wasn't swimming upstream, Jackson liked to relax in the Pacific Ocean with a few friends and a bellyful of plankton, his favourite food.

STEPPE BISON

SADIE
ALASKA

Sadie ran the Ice Age's toughest bootcamp, teaching her fitness regime throughout the mammoth steppe that stretched across Europe, Central Asia, Beringia (the land that once linked Canada and Russia) and North America.

Although Sadie looked quite similar to the bison you might see today, at 4.5 m tall she was much, much larger, with long back legs, huge curved horns and a big hump on her back.

A LION'S TALE

Sadie's name was on everyone's lips after she took on a hungry Beringian lion... and *won*. Her cousin Babe wasn't so lucky. In 1979, 36,000 years after Babe's death, her body was found in an Alaskan gold mine. The minerals in the mine had turned her skin blue. The humans who found her named her Blue Babe.

MODEL MATERIAL

Did you know some of the earliest examples of human art were inspired by steppe bison? Paintings have been discovered in the Cave of Altamira in Spain and the Lascaux caves in France.

DYING OUT

Sadie lived to a grand old age, but other steppe bison weren't so lucky. They were hunted for food by several predators – including Neanderthals. Overhunting could be one reason that they became extinct.

GIANT SLOTH
LAID-BACK LUIS
SOUTH AMERICA

Can you roll your tongue? Around 1.9 million years ago, Luis was the talk of the jungle for the shapes he could make with his long, flexible tongue. And it was good for more than just the Guiness Book of Records. Luis's tongue was perfect for gripping berries and picking leaves.

Instead of lumbering slowly along the ground, Luis dreamed of relaxing in a tree among the branches, leaves and berries. He'd be glad to know that his modern relatives, the moss-covered sloths, have made it into the trees. And they're just as laid-back.

HUNGRY HUMANS
Luis didn't like fighting, but that didn't stop Neanderthals from hurling their spears at him. Although they wanted a quiet life, giant sloths like Luis were easy prey for Neanderthals, who stalked and hunted them for food.

EATING MACHINE

Luis was always hungry and he was famously good at getting food. He would stand on his back legs and hook his long claws around tall branches to pull them down to his mouth.

LOW PROFILE

At the size of an elephant, Luis was too big to live in a tree. In his youth his dreams of tree-top living led him to clamber onto a low branch, but it was a crashingly giant disaster...! Luckily Luis's thick, shaggy fur broke the fall.

ARCTIC GROUND SQUIRRELS
THE FURRY FAN CLUB
RUSSIA

COLD WINTERS
During hibernation, Arctic ground squirrels drop their body temperature to –2.9°C. In the spring, they emerge from their burrows and eat the seeds and grasses they had hidden the previous autumn, as well as fresh mushrooms and insects.

SQUIRREL STYLE
These super-fans have always known how to turn heads, rocking red or yellow fur in summer and then a brand-new silvery look for autumn.

The furry fan club met 10 million years ago while trying to get Gavin the Giant Ape's autograph and they have been star-seeking ever since. These squirrels are proof that you don't have to be big to be successful. They're just 39 cm tall and, centuries after my family became extinct, they're still celebrity-spotting across the Arctic Circle and Northern Hempishere. That's one long celebrity crush!

TUNNEL BUILDERS

The squirrels lived together in groups of up to 50 and they survived the long winters by hibernating in underground tunnel networks for 7 to 8 months. If there were awards ceremonies for sleeping, these guys would have won every prize.

DIRE WOLF
THE BRAT PACK
CANADA

The fearsome Brat Pack caused trouble all the way from modern-day Alberta, Canada, to Bolivia. This gang were pack hunters and no territory – not even grasslands, forests or savannah – proved too challenging. Their hunting techniques were so successful that the modern-day timber wolves still hunt in teams to bring down much bigger prey.

PERFECT PREDATOR

Dire wolves had a short, broad head and powerful jaws that made even me tremble. They were the largest species of wolf ever found, and their big, meat-tearing teeth had me thundering in the opposite direction. I knew they ate horses and bison, but I had a nasty feeling that I might also have been on the menu.

YUKON HORSE
BETTY
CANADA

Betty was the belle of the Yukon with her flowing, blond mane and her white winter coat. She spent her days having fun on the cold steppe with her wide circle of friends and her huge family. Her descendants are modern wild and tame horses.

FAMILY MATTERS
Just like wild horses today, Yukon horses would have lived in large families. Betty's herd would have had a stallion, who was the head of the family, and lots of other mares and foals.

WORLD RECORD
Did you know the oldest DNA ever analysed in the world was from a 700,000-year-old Yukon horse fossil?

GIANT SHORT-FACED BEAR

MARIA

MEXICO

I'm sorry to say that Maria was a bit of a bully. She never bothered me but like all bullies, she only picked on smaller creatures. Even though Maria *could* catch her own food, she preferred to wait until other animals had hunted their prey. Then she just stole it from them, making herself look fierce by standing on her long, muscly back legs – she was 4.3 m tall with her arms up and could scare almost anyone into giving her their lunch.

MIGHTY APPETITE

Now, I'm not making excuses for what she did, but Maria was 900 kg and needed to eat at least 16 kg of meat every day to survive. That was a lot of food to find. Luckily, her long nose could sniff out dead animals from many kilometres away.

ROAMING FREE

Maria stole her meals all across the high grasslands of western North America, from Mexico to Alaska and Yukon, and she could walk for days without needing to rest.

GOOD POINTS

We forgave Maria for her bullying ways after she scared off a pack of curious Neanderthals. These early explorers ran for their lives after they saw her patrolling the Bering Strait (the only entrance to North America).

GIANT TERATORN

PAULA

ARGENTINA

Paula was the stunt superstar of the Late Miocene period. As one of the largest flying birds the world has ever known, she had a wingspan of 7 m – about the size of a small aeroplane and twice the wingspan of an albatross.

CRANKY CREATURE

The name 'teratorn' means 'monster bird' and Paula lived up to her name. She would swoop down to catch small animals like mice and lizards and swallow them whole.

RECORD BREAKER

Getting a body as big as Paula's airborne wasn't easy. To take off she ran down a slope and used winds blowing towards her to help her get into the air, just like a modern-day hang-glider. Once in the air, she could glide all the way from the Andes to the Pampas lowlands in Argentina.

GRACEFUL GLIDER

Teratorns could soar for great distances, moving their wings and using their 150 cm-long flight feathers to control the glide while they searched for prey.

GIANT ECHIDNA

MATILDA

WESTERN AUSTRALIA

Matilda was a self-defence guru and animals travelled from across the world to visit her in her home in the Australian outback to learn her amazing techniques.

The fearsome spines on Matilda's back were part of her favourite self-defence technique. She burrowed her strong limbs into the ground and curled her body into a ball, so only her spiny back was showing. Once she was in a ball there was no moving her. Predators soon got bored of waiting for Matilda to move.

TERRIFIC TONGUE
Worms, insects, grubs and ants were high on Matilda's list of favourite snacks. She used her 50 cm tongue to gobble up whole ant colonies in one meal. What a tongue!

FEELING SHEEPISH
Matilda weighed about 30 kg and was the size of a modern-day sheep. She could stand on her very long back legs so that she could use her huge claws to dig out termite nests.

PUGGLE POWER
Matilda developed her self-defence tools when she was a young puggle, the name for a baby echidna. She and her brothers and sisters could only live in their mum's pouch until their spines started to grow.

35

WOOLLY RHINO

HECTOR

EUROPE

Anyone who was anyone knew Hector, and we all adored him. His love of nature inspired generations of furry rhino babies to follow in his footsteps and teach other animals about the wildlife around them.

Like me, this gigantic plant eater was always dressed for success in freezing temperatures. He had short legs and small ears, which kept them less exposed to the cold. His shaggy fur coat was thick and waterproofed with wax and oil.

HOME SWEET HOME
Hector's old stomping ground was the cold grasslands around the edge of the Arctic in Northern Europe and Siberia. He spent his days searching for new species of lichen and moss, which he had to sample of course!

SENSE-IBLE
Woolly rhinos were very short-sighted but had a great sense of smell. They also had two nose horns made from a mass of hairs that they used to dig up food from under the snow. Just like how I used my tusks!

CENTRAL HEATING
Hector's tummy contained some friendly bacteria to help him digest grass and leaves. As the food was broken down, his tummy created lots of heat, which kept him warm from the inside.

HEADS UP!
It's rare to find the remains of woolly rhinos these days, but a few well-preserved skulls have been discovered at the foot of the Himalayan mountains. They probably froze soon after their death, which is why they weren't eaten by scavengers.

IRISH DEER
CONOR
IRELAND

Conor was such a joker. When he first broke onto the Irish scene 1.5 million years ago, he pretended to be an elk. We couldn't believe it when we found out that he was actually a huge deer!

At 2.1 m tall and 600 kg, Conor had great stage presence. He captivated everyone with his jokes – and he loved to laugh. He'd find it hilarious that these days, his amazing antlers decorate the walls of an Irish castle!

BUSY BOGS
The bogs and lake beds of Ireland are great at preserving bones, thanks to the mix of acidic water, low temperature and no oxygen. That's why so many fossils of Irish deer have been found in them.

THE LONG GOODBYE

When the climate changed, so did Conor's habitat. Grass plains became forests, and Conor's huge antlers kept getting tangled in the leafy branches. Slowly, Irish deer began to disappear.*

A POINTY IMPRESSION

An Irish deer's antlers weighed 40 kg and from tip to tip were twice the length of your bed! They grew a new pair every year – just to show off to female deer.

*Pssst! Rumour has it that Conor's relatives were living in Siberia as recently as 6,500 years ago.

39

GIANT BEAVER

BILL & BEN
CANADA

Bill and Ben were some of the biggest sporting stars of the Ice Age, and their syncronized swimming skills drew crowds of fans after they won the gold medal at the Pleistocene Olympics.

At 2.5 m and weighing 100 kg, Bill and Ben were pretty clumsy on their feet, but in the water it was a different story. They were surprisingly graceful and agile for rodents that were the size of a modern-day black bear.

FANCY FEET
Giant beavers had flat, narrow tails and large hind feet to help propel themselves through the water with speed and ease.

ONE-TRACK MIND

Giant beavers weren't interested in anything except swimming. The beaver craze for building dams came after their time. Can you imagine building a dam big enough for a bear-sized beaver?

BIG GNASHERS

Like many Ice Age celebrities (me included), Bill and Ben were herbivores. They had beaming smiles, with amazing teeth that were perfect for grinding up plants. Their cutting teeth were 15 cm long, with thin ridges and rounded points.

41

MEET THE NEANDERTHALS

My star-studded guide to the Ice Age wouldn't be complete without these guys. They were the talk of Europe and Asia between 400,000 and 40,000 years ago. Neanderthals were a different species of human from you but somewhere way, way back in your family tree, you share a relation.

SMARTY PANTS
Don't listen to the rumours that Neanderthals were stupid. They were clever and adaptable folk who made tools, jewellery, fire and even painted art on the walls of their cave homes. Believe me, they were skilled hunters too!

Bulging brow ridge and larger, flatter heads.

Large, wide nose better for breathing in and warming up the colder air.

Big front teeth that they used as tools to bite and hold on to things.

Shorter bodies better adapted to colder weather as less skin was exposed to the cold.

Stocky lower legs and arms for ambushing their prey during hunts.

43

LA BREA TAR PITS

LOS ANGELES, USA

Back in my day, the La Brea tar pits were a dangerous place. Many careless animals got stuck in the natural asphalt that bubbled up from underground and there was no escape. It took months for bodies to sink below the surface and in the meantime scavengers would feed on them and get stuck too.

Not all the finds at La Brea are big animals. Fossils of pollen, bees and dragonflies have also been discovered. Scientists can learn more detailed information about the climate, ecosystem and food chain from these small organisms.

HOLLYWOOD

Today, the tar pits are famous for having one of the biggest collections of Ice Age fossils in the world. The asphalt has preserved fossils of over 600 species including snakes, sloths, mountain lions and more than 200,000 specimens of dire wolf.

The bones have been kept in excellent condition thanks to the sticky asphalt. When the tar has been very carefully cleaned off the bones, scientists can see tiny details like teeth marks.

45

ICE AGE WORDS

Here's all the lingo you need to know to sound like an Ice Age expert.

ANCESTOR – an early animal or plant from which a later type has evolved.

ARMOURED SKIN – scaly or bony skin that protects an animal from being attacked by its predators.

CARNIVORE – any creature that eats another creature.

CLONING – the scientific process of creating an identical copy of a living thing.

DESCENDANTS – an animal or person related to someone from an earlier generation.

DIGESTION – when your body breaks down food and changes it so you can use it for energy.

EVOLVE – when the body parts of an animal or plant slowly change over time.

EXTINCT – an animal is extinct when no members of its species are alive.

FOSSILS – the mineralized bones or the mark left behind of an animal or plant from an earlier period in time.

GREENHOUSE GASES – gases in the Earth's atmosphere (like carbon dioxide) that trap energy from the sun.

HERBIVORE – an animal that eats only plants.

HERD – a group of animals that lives and feeds together.

HIBERNATION – when animals hibernate, they spend the winter in a deep sleep.

HOLOCENE EPOCH – the period of time that began at the end of the Pleistocene and continues up till the present.

LATE MIOCENE PERIOD – the time period that began about 12 million years ago and lasted until about 5.3 million years ago.

NATURAL ASPHALT – a black, sticky substance that rises to the Earth's surface forming tar pits.

PLEISTOCENE EPOCH – the time period that began about 2.6 million years ago and lasted until about 11,700 years ago.

PREDATOR – an animal that hunts and kills other animals for food.

PREY – an animal that is hunted or caught by other animals for food.

REPTILE – a cold-blooded animal that is covered in scales or plates, and lays eggs.

WINGSPAN – the distance from the tip of one wing to the tip of the other wing.

INDEX

First published in the United Kingdom in 2020 by
Thames & Hudson Ltd, 181A High Holborn, London WC1V 7QX

Text by Rachel Elliot
Designed by Emily Sear

British Library Cataloguing-in-Publication Data
A catalogue record for this book is available from the British Library

ISBN 978-0-500-65206-0

Printed and bound in China by C & C Offset Printing Co. Ltd

To find out about all our publications, please visit
www.thamesandhudson.com. There you can subscribe
to our e-newsletter, browse or download our current
catalogue, and buy any titles that are in print.